Environmental Aspects, Applications and Implications of Nanomaterials and Nanotechnology 2016

Topical Conference at the 2016 AIChE Annual Meeting

San Francisco, California, USA
13 - 18 November 2016

ISBN: 978-1-5108-3426-2

Printed from e-media with permission by:

Curran Associates, Inc.
57 Morehouse Lane
Red Hook, NY 12571

Some format issues inherent in the e-media version may also appear in this print version.

Copyright© (2016) by AIChE
All rights reserved.

Printed by Curran Associates, Inc. (2017)

For permission requests, please contact AIChE
at the address below.

AIChE
120 Wall Street, FL 23
New York, NY 10005-4020

Phone: (800) 242-4363
Fax: (203) 775-5177

www.aiche.org

Additional copies of this publication are available from:

Curran Associates, Inc.
57 Morehouse Lane
Red Hook, NY 12571 USA
Phone: 845-758-0400
Fax: 845-758-2633
Email: curran@proceedings.com
Web: www.proceedings.com

TABLE OF CONTENTS

472319 Invited Presentation: Carbonaceous Nanoparticles and Their Interactions with Biological Cells

Abstract:

Particles originating from human activities have existed for millennia, e.g., smoke from combustion, but the recent development of industry and combustion-based engine transportation has increased anthropogenic particles pollution. At the same time, technological advancement has also changed the character of these particles, increasing the proportion of nanometer-sized particles --"nanoparticles"-- and expanding the variety of chemical compositions. Indeed, the manipulation of matter at the scale of atoms, "nanotechnology," is creating many new materials with characteristics not always easily predicted from current knowledge.

In this work we report on the interactions of carbon-based nanomaterials (both from combustion and synthetic sources) with biological cells, using atomistic simulations. As engineered nanomaterials, we will focus on carbonaceous quantum dots, which have recently emerged and ignited tremendous research interest. Their favorable characteristics include size- and wavelength-dependent luminescence, resistance to photobleaching, bio-conjugation, and functionalization to produce chiral nanostructures. Carbon-based quantum dots show promise in areas such optoelectronics, catalysis, bioanalysis and drug delivery.

Copyright © American Institute of Chemical Engineers. All rights reserved.

463630 Nanomaterials: Occupational Exposure and Toxicity in Context

Abstract:

At the forefront of particulate health concerns are nanomaterials. This talk will propose some considerations for evaluating nanomaterials and their health effects.

1. Is the material as prepared truly â€œnanoâ€? For example, even though SWNTs satisfy the accepted definition of being nanoscale in 2-dimensions, in practice and certainly as synthesized, they are highly coalesced into aggregates of micron in size.

2. What distinguishing characteristics apply to the particular nanomaterial? High surface area is a universal hallmark of a nanomaterial. Yet fumed silica with a surface area of > 200 m^2/g is rated biologically safe and even used a food additive.

3. What are the synthesis methods? Even if synthesis is conducted within a closed system, subsequent processing steps of harvesting and processing can create far greater exposure risk.

3. Do the current markets and/or applications warrant concern? For example quantum dots contain heavy metals, yet they are a commercial product used in limited quantities for biological imaging.

4. What are the exposure routes and levels likely to be encountered in the manufacture, product development and consumer use of nanomaterial based products? Examples in which the nanomaterial, manufacturing process and product are generally considered safe include nanoclays used in various commercial polymers, TiO_2 in sunscreen and carbon black in car tires.

5. Is there any relevant information on related materials and/or nanoparticles and if so, is it applicable?

Examples of nanomaterials and synthesis conditions will be used to address these questions.

The talk will balance these considerations with those nanomaterials derived accidentally such from combustion-based stationary power generation or transportation. For the general population these sources overwhelm their potential exposures to nanomaterials, even after decades of study and emission controls. In fact recent studies have linked soot with asthma, heart arrhythmias and even death.

Copyright © American Institute of Chemical Engineers. All rights reserved.

465578 Dose-Dependent Intracellular Reactive Oxygen and Nitrogen Species Production from Particulate Matter Exposure: Results from Ambient Samples and Chamber Experiments

Abstract:

Epidemiological studies have found associations between elevated particulate matter (PM) concentrations and increased incidences of cardiovascular and respiratory diseases. PM-induced oxidative stress has been suggests as a possible mechanism leading to these health effects. Here we present chemical and cellular oxidant measurements from ambient PM samples and controlled chamber experiments. We investigated reactive oxygen and nitrogen species (ROS/RNS) production from ambient PM samples and laboratory-generated secondary organic aerosols (SOA). Ambient samples (n = 104) were collected during multiple seasons from rural and urban sites around the greater Atlanta area as part of the Southeastern Center for Air Pollution and Epidemiology (SCAPE). SOA were generated from the photooxidation of six commonly emitted volatile organic compounds (VOCs). These experiments were conducted in the Georgia Tech Environmental (GTEC) facility under both dry and humid conditions in the presence of ammonium sulfate seed using H_2O_2 as an OH radical precursor. Murine alveolar macrophages were used to measure intracellular ROS/RNS production, while dithiothreitol (DTT) was used to quantify the concentration of redox-active species in each sample. For each sample, intracellular ROS/RNS was characterized using response parameters obtained by fitting dose-response data using the Hill equation. These response parameters (maximum response, EC_{50}, Hill slope, threshold, and area under the dose-response curve) were compared with DTT activity to determine whether chemical assays represent cellular ROS/RNS responses and with water-soluble PM components to elucidate species associated with PM-induced ROS/RNS production. From both ambient and chamber filters, we found that ROS/RNS production was highly dose-dependent, non-linear, and could not be represented by a single concentration measurement. We demonstrate that there is no simple correlation between DTT activity/PM composition and cellular ROS/RNS responses. For summer ambient samples, we found that DTT activity and organic species were significantly correlated with cellular ROS/RNS production, which motivates further studies involving PM-induced ROS/RNS production from summertime photochemically driven SOA. Finally, from chamber filters, we found that SOA generated from different precursor VOCs contained different concentrations of redox-active species and induced different levels of cellular ROS/RNS.

Copyright © American Institute of Chemical Engineers. All rights reserved.

462217 Biofuel and Reference Diesel Particles: Differences in Inflammatory and Oxidative Effects

Abstract:

Particulate matter (PM) from engine exhaust has become an increasing concern due to its links to numerous adverse effects on human health and the environment. Recent efforts to reduce PM emissions and the associated climate change impacts from combustion include new technologies, such as the installation of diesel particulate filters and the use of biofuels. However, studies comparing the biological effects of PM from biodiesel and petroleum-based diesel provide conflicting results, and interpretation of these results is difficult because of different methods for particle generation and for biological assays.

This study aims to systematically evaluate the effect of particles generated from two biofuel blends, one biodiesel (BD) and one alcohol diesel (AD) and one reference standard diesel particle sample (NIST SRM 2975) on selected biological outcomes associated with responses to diesel particles in lung cells and in animals. These include: (a) CYP1A1 and 1B1 enzymes that can both detoxify and bioactivate PAHs in human airway epithelial cells, human bronchial epithelial cells and human macrophage cells lines, and (b) IL-8 and TNF-a that play critical roles in pulmonary host-defense and inflammatory responses. We also examined the role of activation of the transient receptor potential ankyrin-1 (TRPA1) cation channel, which is a target for diesel exhaust and similar combustion-derived particles and mediator of pro-inflammatory and irritant responses. Because PM deposits in small airways and the alveolar space, cell lines derived from small airway epithelial cells (H441) and modeling alveolar macrophages (THP-1) were selected for this study. PM was generated from simple surrogate fuels and characterized under well-controlled combustion conditions.

The results show that particle extracts caused similar responses compared to whole particles in both upper airway epithelial and macrophage cells. This suggests that the soluble organic components adsorbed on the particle surface are highly associated with the cellular responses we examined. When comparing the particles from the different fuel sources, BD particles are capable of activating key enzymes involved in PAH metabolism to very high levels compared to RD and AD. BD particles also had the greatest surface area, soluble organic fraction and oxidative potential. In contrast, RD was more pro-inflammatory than BD or AD, an effect that involved TRPA1 activation. This study adds to the growing body of evidence that biodiesel PM results in adverse lung health effects, potentially via different pathways than conventional diesel PM.

Copyright © American Institute of Chemical Engineers. All rights reserved.

470397 Real Life PM Emissions from Traffic and Human Exposure Implications

Abstract:

The contribution of vehicle emissions to urban air pollution has been extensively investigated. Such emissions have been found to account for almost 90% of the total amount of substances released into urban ambient air. The ability of solid microparticles to penetrate deep into the human respiratory tract results in increased morbidity and mortality. There is also evidence that car exhaust gases are directly involved in the pathogenesis of several health endpoints, including respiratory, cardiovascular, neurological, and allergic diseases, as well as asthma. With regard to smaller particles that penetrate deeply across human respiratory tract, we need to highlight that particle translocation into systemic circulation is one of the major mechanisms related to air pollution attributed cardiovascular disease. Taking into account all the above, regulatory frameworks have been developed worldwide. Obviously, during the operation of a car, the emission control technologies efficiency declines with mileage, resulting in increased exhaust emissions; however, emissions of conditionally new cars that have a mileage of up to 1,000 km are not studied and unfortunately, not regulated. Considering all the above, the aim of this study is to characterize particulate (segregated by size and active surface) released by cars with low mileage and compare the results with those obtained from in-use vehicles with a mileage of 60,000-280,000 km, as well as to investigate the potential implications to human exposure.

The cars with high mileage and the ones with low mileage were kindly provided by the author colleagues. Conditional codes were introduced for the cars so as to avoid specifying the car manufacturers and models, covering various individual vehicle characteristics including manufacturing year, engine capacity and fuel type. For the aims of the study, the exhaust gas suspension (EGS) method was selected. The measurement methods were patented and used according to the following plan:

1. The test car was started and ran for 1-3 minutes before all extraneous dust and soot particles that had

settled there during the downtime and brought from the outside in the tailpipe were removed.

2. Then, the engine was shut off and the hose was connected to the tailpipe of the test car. After that, the hose was dipped into the plastic container, which was filled with 10 liters of deionized water.

3. After that, the car was started and worked at the neutral speed for 20 minutes. This time interval was chosen considering that the engine had been warmed up for approximately ten minutes (according to a coolant temperature sensor). Then, the warmed up engine idled for the remaining ten minutes.

Once the measurements had been completed, the distilled water container through which the exhaust gases had passed, was sealed with a lid and then sent to the laboratory. A significant advantage of this method is the lack of transformation of aerosols and solid particles. To measure particle size distribution, a laser particle size analyzer supplied with Fritch MaS software was employed. PM size distribution of the wet or dry dispersion units can be measured separately or simultaneously, with automated switching features. Optimal dispersion was accomplished in the NanoTec integral wet dispersion unit by using a combination of a robust, variable speed centrifugal pump with powerful ultrasonification.

To translate the different emission profiles identified in the study into actual exposure and uptake, taking also into account the deposition across HRT, a comprehensive methodological scheme has been developed, taking into account:

a. The amount of PM emitted per km travelled by the vehicles investigated, using the COPERT 4 methodology. PM emission factors (in mass of PM per km traveled) for the different vehicles is necessary so as to account of the overall mass of PM emitted by each individual vehicle type.

b. The PM size distribution of the emitted PM, as identified by the laser particle size analyzer.

c. The intake fraction (IF) of the emitted PM. Intake fraction is a metric of the emission-to-inhalation relationship, facilitating comparisons among sources (in our case the individual vehicles) in terms of their exposure potential. For a given emission source and pollutant, intake fraction is the cumulative mass inhaled by the exposed population divided by the cumulative emissions. Considering that IF depends on several parameters affecting the emission-to-intake process, (e.g. prevalent wind, emissions strength, population density), it is expected that it would vary with location and time. In a recent global intraurban intake fraction study, it was found that population-weighted mean intra-urban intake fraction for cities is equal to 32.

d. The deposition of PM across the HRT. To better understand the exposure implications of size-segregated particles distribution emitted by the investigated vehicles, HRT particle deposition modeling was applied to reckon the PM fraction deposited (DF) to the three parts of HRT. Major mechanisms of PM deposition across HRT include diffusion, sedimentation and impaction. Secondary mechanisms involve interception and electrostatic deposition. Different HRT regions involve different deposition mechanism as a function of PM size as follows:

- Naso-pharyngeal region (or upper respiratory tract – URT): impaction, sedimentation, electrostatic (particles > 1 μm)

- Tracheo-bronchial (TB) region: impaction, sedimentation, diffusion (particles < 1 μm)

- Pulmonary (P) region: sedimentation, diffusion (particles < 0.1 μm)

Several parameters that affect HRT deposition have been taken into account, including PM properties (concentration and size distribution), air flow parameters (lung capacity and breathing frequency) and HRT physiology (structure and morphology). HRT deposition was carried out using the Multiple Path Particle Deposition (MPPD) v. 2.1 model. Software inputs include morphological parameters of the pulmonary system – functional residual volume (FRC), tidal volume (TV), upper respiratory tract (URT) volume, as well as breathing frequency (BF) for each age group.

The results of the study indicated that the mean diameter of particulate emitted from high mileage cars ranges in hundreds of microns. Smaller particles were found in cars that were manufactured over 10 years ago and have considerable mileage. Thus, particles were detected in five of the 17 cars. The same percentage of cars (30%, 5 out of 17) is the source of PM50 class. Particles with a mean diameter of more than 100 μm were found in most samples of the cars without mileage (58%, 10 out of 17), but 42% of the cars without mileage are a source of particles of up to 20 μm. It appears that these aerosols correspond to different agglomeration stages. Data on the surface area of the exhaust gas particulate showed that entirely new cars emit extremely small particles with a large surface area (up to ca. 90,000 cm^2/cm^3).

The qualitative analysis of the exhaust gas particulates showed that there are limited differences in their composition between cars with high and low mileage respectively.

In terms of implications for population exposure and public health, age-dependent physiological differences are also of special importance. Taking into account the vehicle emission profile (mass of PM emitted per km travelled and PM size distribution), the respective intake fraction and the HRT deposition based on the size distribution, the amount of PM retained by all HRT regions was estimated. Among the newest vehicles, the highest amount of PM retained is associated to the emissions of the diesel vehicle VT 2012; the very low mean diameter of the emitted PM implies that its vast majority is within the respirable size spectrum. In general, among the newer vehicles, 0.0096 μg were deposited per km travelled for the diesel vehicles, while the respective amount for the gasoline cars was 0.0074 μg. Regarding older cars, gasoline cars contribution was the same to the newer ones (0.0074 μg per km traveled), while for diesel vehicles the amount (0.031 μg per km traveled) was the highest among all investigated vehicle categories. This observation is attributed to the higher amount of overall PM emitted (per km travelled) by older vehicles. Considering that newer diesel vehicles emit PM of lower aerodynamic diameter compared to older diesel vehicles, a higher fraction of the PM taken up is expected to deposit to the lower HRT.

Among the investigated vehicles, the diesel vehicle encoded VT 2012 combines all characteristics that may be potentially harmful for public health; the lowest average PM diameter (3.84 μm) that results in significant PM deposition across HRT (and especially the lower part), the highest active surface (89,871 cm^2/cm^3), and the highest adsorption potential of organic (e.g. PAHs) and metals.

A key finding of the study is that car mileage has virtually no effect on the size of the solid particles released into the environment, and that even cars with very low mileage (up to 1,000 km) are not "clean" in terms of emissions. On the contrary, special attention has to be paid to the lower aerodynamic diameter related to newer diesel vehicles, their higher specific surface (as well as the higher capacity of toxic substances adsorption) and how this is translated into actual increased human exposure and, consequently, health risk.

Copyright © American Institute of Chemical Engineers. All rights reserved.

467482 Disinfection of Water Using Silver and Copper Nanoparticle Impregnated Activated Carbon

Abstract:

Access to safe drinking water is one of the biggest challenges across the globe. In this context, activated carbon is widely used as a water purification media for producing drinking water, which although removing most of the contaminants (responsible for taste, colour, odour, organics, heavy metals etc.), however is incapable of killing pathogens. To mitigate this, antibacterial properties of silver and copper nanoparticles (Ag-NPs and Cu-NPs) have been utilized. Therefore, the objective of this research is to combine both purification and disinfection effects of activated carbon and nanoparticles together and thereby produce drinking water in a continuous, flow-column set-up.

To this end, first we synthesized Ag-NPs by UV reduction of aqueous silver nitrate solution, in presence of tri-sodium citrate. The latter acts both as a reducing and a stabilizing agent. Following this, Ag-NPs were selectively impregnated on the external surface of plasma treated AC granules. This hybrid is referred to as Ag-AC hybrid, having 0.8 wt% of Ag-NPs of 28 nm mean size. Subsequently, microbial cell-killing experiments were performed in two set-ups: (i) shake flask (called batch experiments) and flow-column (called continuous-mode). We have used *E. coli* K12 (MTCC 1302) as a model organism (as it is typical of contaminated drinking water), always at a concentration of 10^4 CFU cells/ml in the inlet water. Batch results using 8 mg Ag-AC hybrid/ml of cell suspension showed that, all cells were killed within 25 minutes of contact time. Maintaining same *E. coli* concentration and Ag-NP loading percentage in Ag-AC (as in the batch experiment), a glass column (8 cm diameter and 25 cm in height) packed with Ag-AC was used for monitoring cell-killing performance over a long duration. Again, inlet water having 10^4 CFU of *E. coli*/ml could be completely disinfected to produce treated, outlet water having zero cell count, with a contact time of 23 minutes, very close to the designed value based on the previous batch cell-killing kinetics. Therefore, a continuous throughput of treating 1.62 L of contaminated water per hour could be maintained for at least up to 16 days, beyond which experiments were not conducted presently. Moreover, the Ag concentration in the outlet water reached only up to a maximum of 29.8 μg/L at steady state, which is well within the recommended maximum Ag concentration limit of 100 μg/L for drinking water (EPA). Thus these results demonstrate the viability of such a column-design to produce decontaminated water at a reasonable rate per hour, over a long period and with minimal loss of Ag in the treated water.

Cu-NPs being reported for its antibacterial activity too, are actually cheaper to make than Ag-NPs. For some of the Ag-resistant microorganisms, Cu-NP also shows better activity compared to Ag-NP. Thus, Cu-NPs have been synthesized using the thermal reduction method where L-ascorbic acid reacts with cupric chloride dihydrate to form Cu-NPs. In this green route, L-ascorbic acid works as a reducing agent. Cu-NPs of mean diameter 5.17 nm were synthesized and subsequently impregnated on plasma treated AC by wet impregnation. The Cu-AC showed a copper loading of 2.1 wt. %. Disinfection studies were performed using these Cu-AC, again in both batch and continuous modes, using the same wild type model microorganism *E. coli* K12 (MTCC 1302), starting with the same initial cell concentration of 10^4 CFU/ml. Similar to Ag-AC, on using 8 mg/ml of Cu-AC in the batch experiment, zero cell count was achieved in only 10 minutes. Based on this batch mode cell-killing data, the same glass column (but packed with only Cu-AC now) was used for continuous, flow experiments. When 10^4 CFU/ml of *E. coli*cells were introduced in the inlet water to this column, zero cell count in the outlet water was achieved in a contact time of 9 minutes, giving a maximum treated water flow rate of 3.77 L/h. Therefore, compared to 1.62 L/h of water in case of Ag-AC, a much higher amount of 3.77 L/h of water could be produced in the same column with Cu-AC. Simultaneously, in steady state continuous flow-column experiments, the maximum copper concentration in the outlet water was around 83.4 μg/L, which is far below the upper recommended limit of 1000 μg/L (EPA). Thus, Cu-AC shows better cell-killing performance compared to Ag-AC. This is primarily because the Cu loading (2.1 wt%) is higher in Cu-AC, compared to Ag loading of 0.8 wt.% in Ag-AC. Moreover, the steady state release concentration of Cu was much higher (83.4 μg/L) compared to Ag release concentration (29.8 μg/L) from Ag-AC. Therefore, Cu-AC shows higher antibacterial activity compared to Ag-AC and reaches zero cell count within a shorter contact time of 9 minutes in the continuous flow-column, in contrast to 23 minutes required for Ag-AC column to achieve zero cell count.

As Cu-AC shows better antibacterial activity compared to Ag-AC, further experiments are being pursued now with a mixture of Ag-AC and Cu-AC. Both these samples are mixed at 1:1 wt. ratio, that is, Ag-AC having 0.8 wt% Ag loading and Cu-AC having 2.1 wt% Cu loading. These have also been further used in both batch and continuous, flow-column experiments for cell-killing. Starting with the same 10^4 CFU/ml *E. coli*, zero cell count has been achieved in 8 minutes, using 8 mg/ml of the hybrid mixture. This implies that Ag-AC and Cu-AC combination potentially gives a synergistic effect in terms of *E. coli*cell-killing and is hence superior to the usage of only Ag-AC or only Cu-AC. Currently, further continuous column experiments are in progress, using the combination of Ag-AC and Cu-AC to optimize its performance towards the fastest possible cell-killing kinetics, which will maximize the production rate of drinking water.

Hence, water disinfection for potable quality (drinking) water (zero *E. coli* count with < 100 μg/L Ag and < 1000 μg/L Cu) could be achieved in a continuous manner at a large flow rate of nearly 4 L/h over a long duration, by using mixed packing of Ag-AC and Cu-AC.

Copyright © American Institute of Chemical Engineers. All rights reserved.

457443 Metal Organic Framework Composites for Chemical Protection

Abstract:

Several MOFs have shown remarkable capabilities of degrading chemical warfare agents (CWAs) and their simulants. However, the exploration of MOFs in engineered materials such as textiles and clothing for chemical protection has not been prominently explored. It is envisioned that the next generation of protective clothing can shield environmental contaminants by integrating reactive materials, such as MOFs, into the fabric, where in addition to providing a physical barrier, the protective clothing can also undergo self-decontamination. Toward that end, we have engineered a nonwoven fiber composite containing reactive Zr-based MOFs that show the ability to degrade G- and V-type nerve agents. Our approach allows us to load a high weight content of MOFs on the fabric and offers access to reactive sites on the MOFs. We demonstrate through simulant and chemical agent reactions that MOFs in the fabric matrix remain highly reactive and mimic reactions of native MOF powders. These newly developed textiles are further evaluated for breathability, moisture vapor transport, and toxic gas permeation rates across the composite. We use DRIFTS to quantify the interactions between the substrate and reactive sites on the MOF to gain understanding of their reactivity. Our approach will serve as a platform for engineering MOFs into usable textiles without compromising MOF reactivity and performance.

Copyright © American Institute of Chemical Engineers. All rights reserved.

458151 In-Vitro Dosimetry Model for Toxicity Ranking of Metal Oxide Nanoparticles

Abstract:

There is a debate in the nanosafety community regarding whether cellular toxicity of engineered nanoparticles (NPs) should be ranked on the basis of delivered dose (i.e., NP mass settled per suspension volume) or administered dose (i.e., initial NP mass concentration). Calculation of delivered dose with sufficient confidence for in-vitro toxicity ranking of NPs requires quantification of NP sedimentation considering fundamental suspension properties of NPs in cell culture media. Accordingly, we developed a first principles â€œparticles in a boxâ€ sedimentation model, which accounts for the particle size distribution, fractal dimension and permeability of agglomerated NPs. The developed sedimentation model was evaluated against both external and our own experimental sedimentation data for metal-oxide NPs, demonstrating significantly improved estimates of settled NP mass over other simplified sedimentation calculations with the assumption of impermeable and/or monodispersed NP agglomerates. The model was then applied to calculate the delivered NP doses in different cell culture media for the dose-response analyses of seven metal oxide NPs that were identified toxic from a library of twenty-four NPs in our previous studies. Hierarchical hazard ranking of the identified seven toxic NPs, according to the their EC_{50}, was not measurably different when determined on the basis of either delivered or administered doses. In contrast, simplified sedimentation calculations based on the assumption of monodispersed impermeable NP agglomerates led to significant underestimation of the settled NPs mass and thus distorted toxicity ranking.

Copyright © American Institute of Chemical Engineers. All rights reserved.

472232 Amorphous Silicon Dioxide Nanoparticle Interactions with Pulmonary Epithelial Cells with and without a Pre-Existing Protein Corona

Abstract:

Silicon dioxide nanoparticles are increasingly abundant in consumer products and industrial applications, which increases their potential exposure to humans. In addition, these nanoparticles are under investigation for biomedical applications within the body such as targeted drug delivery. In either case, upon entering the body, nanoparticles interact with biological fluids and are immediately coated with proteins. Ultimately, it is this protein layer that interacts with the bodyâ€™s cells and tissues, and thus dictates the subsequent fate of the nanoparticles in the body. In this work, we investigated the adsorption of an abundant protein in biological fluids, albumin, to silicon dioxide nanoparticle surfaces and the impact of this protein layer on subsequent nanoparticle uptake within human cell cultures.

A previous mechanistic study conducted in our group has indicated that pH plays a strong role in the protein affinity for the nanoparticle surface (*publication in preparation*). Using attenuated total reflectance Fourier transform infrared (ATR-FTIR) spectroscopy, we obtained a real-time adsorption profile for BSA on the nanoparticle surface. Protein coverage on the nanoparticle surface was quantified using thermogravimetric analysis. Results from these studies indicated that at the isoelectric point of the BSA-silicon dioxide nanoparticle complex, pH 3.7, protein coverage was highest. Whereas at physiological pH, pH 7.4, much lower adsorption was observed over the same time period.

Cellular responses to uncoated particles and particles coated at these two pHs, to model a low-coating and a high-coating of protein on the nanoparticle surface, were conducted .Two pulmonary epithelial cell lines, 16HBE14o- and A549, were chosen for these studies as models of the different regions of the lung (bronchioles and alveoli) where nanoparticles deposit. Cell viability was measured via the MTT assay, particle uptake through TEM, and inflammatory markers through ELISA. Results from these studies indicated that coated particles behaved differently than uncoated particles in every parameter examined and that uptake depended on the cell system studied.

Overall, exposure to amorphous silicon dioxide nanoparticles induced a cellular response in pulmonary epithelial cells. Since these particles impacted the two cell lines differently, the deposition location of particle in the lungs after inhalation is important for understanding their toxicity. Surface functionalization prior to cellular exposure also played a key role in the cellular response, further supporting the idea that the protein corona affects cellular recognition and responses.

Copyright © American Institute of Chemical Engineers. All rights reserved.

466807 Physiochemical Properties of Nanoparticles Determine Their *In Vitro* Cytotoxicity

Abstract:

Solid nanoparticles have proven useful in overcoming many of the barriers to drug delivery. Additional development has seen the attachment of a myriad of surface coatings such as polyethylene glycol (PEG). This commonly used PEG layer masks particles from innate host defence mechanisms within the body and increases circulation half-life of the therapeutic upon injection. Although this approach is widely accepted, the underlying mechanism is not well understood. This study examined the effect of hydrophilic surface coatings on solid silica nanoparticles with their interactions with Chinese Hamster Ovary (CHO) cells. Specifically, we examined the effect of differing lengths of PEG on intracellular uptake and viability.

In order to elucidate the effects of particle surface characteristics, studies were conducted with PEG 2k, 5k and 20k, as well as aminated silica nanoparticles. Two nanoparticle sizes, 60 and 120 nm, were chosen to draw size effect comparisons between the particles. These particles are generally recognized as a safe material by the Food and Drug Administration and provided a uniform foundation via size distribution, surface characteristics and mild reaction chemistry that precluded any nanoparticle-mediated effects.

In vitro distribution of the silica nanoparticles was found to directly correlate with their surface coating. Pegylated particles were found to remain interspersed among the CHO cells, whereas amine coated particles were found to produce punctate regions of fluorescence on and around the cells. Confocal microscopy confirmed uptake over simple surface association of the nanoparticles. Similar trends were correlated with flow cytometry. It was found that all PEG coatings decreased particle uptake by the CHO cells in comparison to those with an amine coating. The aminated particles had the greatest effect on cell growth. The 60 nm particles yielded greater half maximal inhibitory concentrations (IC_{50}) values over the 120 nm particles. The IC_{50} values for 60 nm aminated, PEG 2k, 5k and 20k particles were found to be 0.42 ± 0.18, 2.58 ± 0.09, 1.75 ± 0.11 and 3.535 ± 0.15 mg/mL, respectively. The IC_{50} values for the 120 nm aminated, PEG 2k, 5k and 20k particles were found to be 0.59 ± 0.05, 1.435 ± 0.13, 1.57 ± 0.125 and 1.333 ± 0.17 mg/mL. Growth inhibition directly correlated with particle concentration regardless of surface character. Pegylated 60 nm particles showed the greatest differences among their PEG coatings whereas the 120 nm pegylated particles showed similar results among all PEG sizes.

This study found that size and surface coatings have a dramatic influence on the cell viability and uptake *in vitro*. In addition, these studies show there may be an interplay between both of these variable sets. This work has provided a basis of comparison that could be extended to various nanoparticle systems. Understanding the impact of widely used solid nanoparticle compositions, sizes, coatings and their combinations on mammalian systems is an important step in developing robust drug delivery vehicles.

Copyright © American Institute of Chemical Engineers. All rights reserved.

471757 The Impact of Titanium Dioxide Nanoparticles on the Lysosome-Autophagy System and Cellular Clearance

Abstract:

Titanium dioxide (TiO_2) is "generally regarded as safe" by the US Food and Drug administration and is found in a wide variety of consumer products, from cosmetics and pharmaceuticals to food additives. Recent studies suggest that the cellular uptake of nano-sized TiO_2 may cause oxidative stress, lysosomal impairment, and inflammation. However, the impact of nano-sized TiO_2 on cell physiology remains unclear. To address this knowledge gap, we evaluated the cellular response to the uptake of anatase TiO_2 nanoparticles (15, 50, and 100 nm in diameter), with particular emphasis on the lysosome-autophagy system, which is the cell's first response to the uptake of a material perceived as foreign and potentially harmful. Following characterization of the TiO_2 nanoparticles, cytotoxicity analyses were conducted and each of the three TiO_2 nanoparticles were found to induce minimal toxicity at concentrations in the cell culture medium up to 1000 μg/mL. Induction of the lysosome-autophagy system was then tested by monitoring the subcellular location of the transcription factor EB, a master regulator of lysosome- and autophagy-related genes. We found TiO_2 nanoparticles to induce activation of the transcription factor EB at concentrations as low as 10 μg/mL and to induce autophagic vesicle formation. However, uptake of high concentrations of TiO_2 nanoparticles (100 μg/mL and above) was also found to prevent vesicle turnover. We hypothesized that this observed effect on vesicle turnover was due to lysosomal impairment. Therefore, we analyzed lysosomal hydrolase function and membrane stability following TiO_2 nanoparticle uptake and observed a correlation between lysosomal function and integrity and vesicle turnover. In conclusion, we established that cellular uptake of TiO_2 nanoparticles induces activation of the lysosome-autophagy system, and thus upregulation of cellular clearance, but results in a blockage of autophagic flux and degradation *via* impairment of the lysosome.

Copyright © American Institute of Chemical Engineers. All rights reserved.

458173 Potential Impact of Sublethal Levels of Nanomaterials on Interactive Behavior of Environmental Bacteria

Abstract:

Although bactericidal effects of a wide range of nanomaterials on environmental bacteria have been extensively reported, little is known about the impact of nanomaterials at sublethal levels on environmental bacteria and their interactive behaviors, which is a critical gap toward a comprehensive understanding of the impacts of nanomaterials on microbial ecosystems. In this study, using model environmental bacteria, we report the impact of single-wall carbon nanotubes (SWCNTs) and nanoSilver (nAg) on cell-to-cell interaction through influencing the production of â €œpublic goodsâ€ or the production of quorum sensing (QS) signals. Sublethal level of SWCNTs (40 or 80 Âµg/ml) inhibited the production of pyoverdine (PVD) (an important metabolite that involves in interactive behavior of microbial communities, especially in iron-limiting environments) in planktonic cultures of *Pseudomonas aeruginosa*. The inhibitory effect has been confirmed at the mRNA level by transcriptional assay and quantitative polymerase chain reaction (qPCR) analysis. We further explored the effect of SWCNTs on mature biofilms of *P. aeruginosa*. Our results showed that the exposure to SWCNTs for up to 48 h did not significantly influence PVD production in cells encased in mature biofilms; however, the exposure to SWCNTs promoted cell detachment from the biofilms and decreased PVD production of the detached cells. To explore the influence of nanomaterials on quorum sensing, we used *P. syringae* and *Pantoea stewartii* as model organisms and quantified the influence of SWCNTs and nAg on the production of different acylated homoserine lactones (AHL) signals. Our results showed for the first time that differential effects could be exhibited by nanomaterials on different QS signals. Our findings advance current knowledge toward a comprehensive understanding of the ecotoxicity of nanomaterials.

Copyright © American Institute of Chemical Engineers. All rights reserved.

464838 Microwell Arrays for Screening Interactions Between Root-Associated Microbes

Abstract:

Plant growth promoting rhizobacteria (PGPR) associate with root surfaces in order to provide benefits to the plant host, some of which include improved nutrient uptake, pathogen protection, and plant hormone regulation. An improved understanding of the relationships between PGPR and root hosts has spurred recent interest in developing constructed, artificial communities of PGPR for use as a low-energy, sustainable alternative to chemical fertilizers. However, wide-spread application of constructed communities has been limited due to the fact that the vast majority of microbe-microbe interactions occurring between PGPR in the rhizosphere are unknown or poorly characterized. This limitation stems from the fact that traditional microbiological assays designed to characterize microbial interactions do so using co-culturing methods that are qualitative and low-throughput. Additionally, these assays fail to re-create the physical and chemical complexities of natural ecosystems, such as soil or rhizospere, which are spatially confined microenvironments. To address these types of limitations, we have developed a microfabricated cell-screening platform termed the â€œmicrobe arrayâ€ which uses a novel biofabrication method to randomly assemble thousands of independent bacterial meta-populations into microscale wells with controlled physicochemical properties. It was recently shown that by tuning well dimensions, interacting bacteria can be combined into wells at low dispersity, allowing for the assembly of replicate initial populations; or at high dispersity, allowing for thousands of unique meta-populations to be assembled across a single array of wells. After this, communities can be simultaneously screened for growth or decay to identify combinations of bacteria and environmental parameters that promote or inhibit colonization. In development of this platform, we investigated the effect of spatial confinement on the growth of *Pseudomonas* bacteria. Here, populations were assembled into wells at high dispersity, and growth screening experiments revealed that successful colonization of microbial communities in a spatially confined, diffusion limited environment is highly dependent on inoculum levels. Our current work focuses on screening rhizobacterium populations in this platform to identify pathogenic, competitive or symbiotic interactions that inform the development of constructed communities for biofertilizer and biocontrol applications.

Copyright © American Institute of Chemical Engineers. All rights reserved.

EFFECT OF SIZE AND CHARGE OF METAL IONS ON HYDROGEN PEROXIDE STABILITY IN SILICA HYDROGELS

Ezgi Melis Dogan, Fulya Sudur Zalluhoglu, Nese Orbey, University of Massachusetts Lowell, Lowell, MA, USA

Abstract

Currently, the most efficient method to increase the stability of hydrogen peroxide (H_2O_2) is microencapsulation in silica hydrogels and xerogels via environmentally-friendly sol-gel chemistry using sodium silicate as the starting precursor. The high stability of H_2O_2 in these systems arises from strong hydrogen bonding between silanol groups (Si-OH) and peroxide molecules. Our previous study showed that the amount of Na^+ ions have a strong effect on the structure and morphology of hydrogels. Therefore in the present work, the effect of the size and charge of the ions other than Na^+ on the properties of the hydrogel and stability of peroxide was studied. Monovalent (K^+) and divalent (Mg^{+2}) ions were introduced into the sodium silicate precursor in the form of their respective salts. The pH of the sol was changed by using an ion exchange resin to replace a portion of the ions in the solution with H^+ ions. Gelation time and structure, pH of the sol and stability of H_2O_2 in hydrogels were investigated. Results showed that the stability and gel time can be adjusted to the desired level by changing the ion content and pH of the sol. The increase in total ion concentration in the sol resulted in a decrease in gel time and the best H_2O_2 stability was obtained with the addition of the magnesium salt: seventy percent of the peroxide was retained in hydrogels containing Mg^{2+} ions, as compared to 61 % retention with hydrogels containing Na^+ ions at the end of 70 days. Hydrogels were characterized by using scanning electron microscopy (SEM) and Brunauer-Emmett-Teller (BET) analysis.

Introduction

Hydrogen peroxide (H_2O_2) is an environmentally friendly oxidant which has a limited stability and tend to decompose into oxygen and water according to following reaction.[1]

$$H_2O_{2(aq)} \rightarrow H_2O_{(l)} + \frac{1}{2}\,O_{2(g)} \qquad (1)$$

The concentration of the peroxide, temperature, pH of the aqueous solution, and presence of impurities or stabilizers significantly affects the decomposition rate of H_2O_2. Storage at low temperatures is required for H_2O_2 solutions due to this instability.[2] However, H_2O_2 stability at ambient temperatures is desired by many industrial applications such as sanitation,[3] disinfection,[4] and tooth whitening.[5]

Various methods have been experimented to decrease the decomposition rate and increase the stability of H_2O_2.[6-10] Among these methods, microencapsulation of H_2O_2 in silica hydrogels and xerogels has been the most efficient methods to enhance peroxide stability because of formation strong hydrogen bonds of H_2O_2 molecules with the silica surface.[11]

The synthesis of hydrogels and xerogels is performed using sol-gel chemistry which is a simple and low cost method.[12] The most common precursor of the gels is sodium silicate solution. A colloidal suspension known as sol is first formed by introducing silanol groups into aqueous sodium silicate and removing sodium ions from the solution. For this purpose, a cation exchange resin is used to adjust the sodium content via ion exchange in the solution.[11] This exchange occurs according to following reaction where Rz is the styrene-divinylbenzene copolymer and SO_3H is the sulfonic acid group.

$$Na^+ + RzSO_3^-H^+ \rightarrow H^+ + (RzSO_3)Na \qquad (2)$$

A hydrogel is observed in the sol-gel transition point where the viscosity of the solution significantly increase. At this point, Si-O-Si bonds are formed via polymerization of silanol groups in the solution resulting a three-dimensional interconnected network (hydrogel). Xerogels are more contracted gel networks obtained by drying the hydrogels. Silica xerogels containing H_2O_2 has been studied by Żegliński et al.,[6] Bednarz et al.[7] and Wolanow et al..[8] Hydrogel and xerogel properties depend on the drying and storage temperatures, pH of silica sols, and initial precursors. Sudur et al.[9, 10] studied the transition conditions of hydrogel to xerogel and observed the effects of sodium ion on the pH of the sol and H_2O_2 (3.3 to 33.2wt %) stability at room temperature (RT). Significant effects of sodium ion content on the structure, the release rate of entrapped H_2O_2, and the thermal properties of hydrogels and xerogels were found.

The effects of size and charge of ions present in the sol on the structure of hydrogels and stability of entrapped H_2O_2 were questioned because of the observations in our previous work.[9] In the literature, the effect of ion type and concentration on the binding power of sodium silicate was studied for applications in petroleum drilling and construction.[13-15] However, H_2O_2 was not incorporated into the formulations used in these studies. To best of our knowledge, the effect of size and charge of ions on the hydrogel formation or hydrogen peroxide stability has not been studied.

In the present work, H_2O_2 was encapsulated in silica hydrogels using K^+ ions into the sodium silicate precursor in the form of potassium chloride (KCl) at three different molar ratios of K^+: Na^+ (0.14:1, 1:1, and 1.76:1). The effects of K^+ on the gelation time, pH of the sol, stability of H_2O_2 and gel structure were studied. Similar to Na^+, K^+ is monovalent but it has a larger atomic size. On the other hand, silica hydrogels were obtained using Mg^{2+} in the precursor solution to study the effect of charge of the ion on H_2O_2 stability.

Experimental Procedure

Materials

Sodium silicate solution (reagent grade, 10.6% Na_2O, 26.5% SiO_2), hydrogen peroxide (30wt % in H_2O, ACS reagent), Amberlite IR120 (hydrogen form), orthophosphoric acid (85% wt in H_2O); potassium permanganate (ACS reagent >99%) and sulfuric acid (99.999%) were purchased from Sigma Aldrich Chemical Co. Potassium chloride, magnesium chloride and calcium chloride (ACS reagent, >99%) were purchased from J.T. Baker.

Hydrogel synthesis

The procedure of hydrogel synthesis, previously available in the literature[9] was modified. Hydrogels containing Mg^{2+} and K^+ ions along with Na^+ ions were synthesized dissolving magnesium chloride ($MgCl_2$) and potassium chloride (KCl) in deionized water before the addition of sodium silicate. Sodium silicate solution was added dropwise to salt solutions at a volume ratio of 1:5. For hydrogels containing K^+, three different K^+: Na^+ molar ratios of 0.14:1 (low), 1:1 (equimolar), and 1.76:1 (excess) were formulated. In hydrogels containing Mg^{2+}, the molar ratio of Mg^{2+}: Na^+ was adjusted as 0.2:1. Then, all formulations are combined with 30wt % H_2O_2 solution and 7 μL of phosphoric acid (H_3PO_4). The pH of the sols were controlled by adding different amounts (0-3.2 g) of ion exchange resin (Amberlite IR 120, hydrogen form) to replace a portion of the ions in the solution with H^+ ions. The mixture was stirred with an overhead mixer for 4 min. The sol was transferred to a glass Petri dish (D = 100 mm, h = 15 mm) after observing the settlement of the resin. The petri dish was placed in a BlueM gravity oven at 40 °C until observing the gelation.

Characterization

The pH of the sol was changed using different amounts of ion exchange resin at a constant K^+: Na^+ ratio and an initial H_2O_2 concentration (19.9wt %). Gelation time was observed with respect to pH value of the solution for each K^+ formulations. H_2O_2 stability was observed in hydrogels containing K^+ as a function of pH and molar ratio of K^+: Na^+ ions. Stability of H_2O_2 in hydrogels containing Mg^+ was compared based on the presence of ion exchange resin in the starting sol. Hydrogels were stored under ambient conditions at room temperature, and H_2O_2 content was determined using titration method with potassium permanganate ($KMnO_4$) as the titrant. Measurements were made daily for the first 10 days and, then periodically until the concentration of H_2O_2 dropped to 2wt %. Two repeated titrations were performed for each sample and the reproducibility was found to be within 1%.

Quantachrome Autosorb-3B nitrogen sorption porosimeter and JEOL JSM 7401F scanning electron microscope were used to study the morphology and structure of the hydrogels. The surface area and pore size were measured through BET (Brunauer-Emmett-Teller) analysis in the porosimeter. Before these measurements and SEM imaging, aerogels were obtained from hydrogels using a critical-point dryer (Tousimis SAMDRI-795) to preserve the solid network and retain the pore structure. Aerogel samples were coated with gold/palladium target by using a vacuum sputter coater. SEM images of samples were obtained at an acceleration voltage of 10 to 15 kV.

Results and Discussion

Gel time as a function of pH and K^+: Na^+ ratio

Gelation time of low and equimolar of K^+: Na^+ ratio in the sol was studied for different pH values of the sol. Results are plotted in Figure 1. Gelation was most rapid between pH=4 and 6 because the sol passed through the isoelectric point, at which it did not carry a net electrical charge and rapid aggregation of the particles was observed.

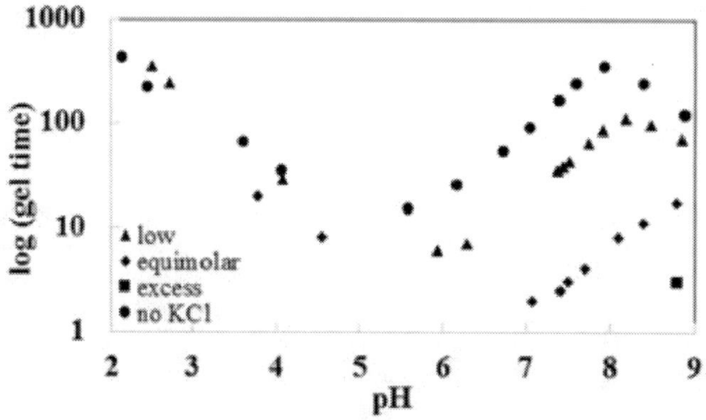

Figure 1: Relationship between gel time and pH for different molar ratios of K^+: Na^+ ions ▲: low (0.14) ♦: equimolar (1) ■: excess (1.76) ●: previous results obtained with unmodified sodium silicate precursor[9]

For all the pH values studied, addition of the ions resulted in faster gelation compared to the previous results obtained in the presence of Na^+ ions only.[6] At the pH range of 2-4, a slow gel formation was resulted because of low concentration of hydroxyl ions. However, longer gel times were obtained after the isoelectric point (pH=6-8) because the gel become negatively charged by high concentration of hydroxyl ions. In the presence of KCl, a shift toward the basic side was observed in Figure 1. This is because the K^+ and Cl^- ions have higher affinity for silicate ions rather than OH^- ions and, therefore more OH^- ions remain in the sol.

Stability of H_2O_2 in hydrogels containing K^+

H_2O_2 content was measured during 70 storage days to study the stability of H_2O_2 as a function of the K^+: Na^+ ratio and the pH of the sol. Stability results of hydrogels containing different K^+: Na^+ ratio at a constant pH of 8.8 are shown in Figure 2. H_2O_2 stability was observed to increase with the addition of K^+ ions as seen more clearly in short term stability (first 10 days during storage).

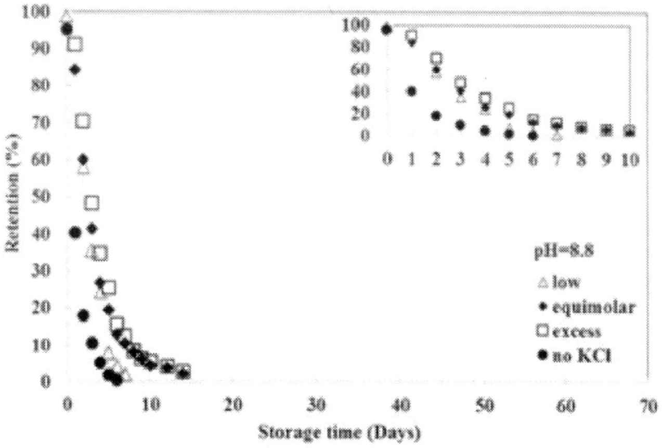

Figure 2: Room temperature stability of H_2O_2 as a function of storage time at constant pH=8.8 (for different molar ratios of K^+: Na^+ Δ: low ♦: equimolar □: excess ●: previous results obtained with unmodified sodium silicate precursor[9])

For example, 18.1wt % peroxide retention was obtained at the end of 2 days of storage at room temperature in our previous work when no K^+ ion was present in the system.[9] In this study, the peroxide retention of hydrogels containing low, equimolar and excess K^+:Na^+ molar ratios was 58.1, 60.2 and 70.6wt %, respectively, during the same period. Moreover, complete decomposition of H_2O_2 in hydrogels containing K^+ took longer (7 to 14 days) than that in previously studied hydrogels[9] (5 days).

Figure 3 a & b show the effects of pH on peroxide stability for low and equimolar amounts of K^+: Na^+, respectively. For each ratio, when the pH of the sol decreased, higher retention of H_2O_2 was observed. At lower pH values of the sol, higher amounts of H^+ ions increase the entrapment efficiency and retention by creating more available sites for peroxide in silica network.[16]

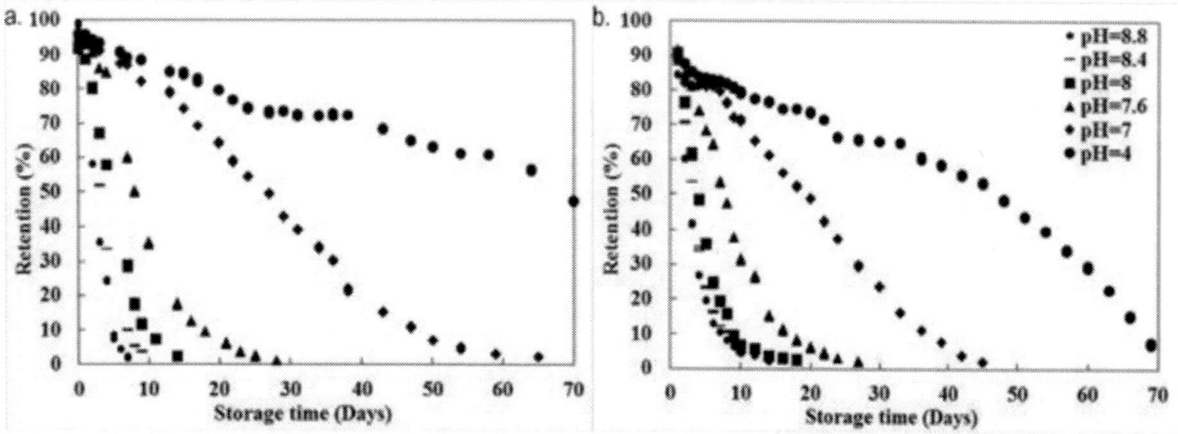

Figure 3. The comparison of H_2O_2 stability in hydrogels containing low (a) and equimolar (b) ratios of K^+: Na^+ at different pH values

Stability of H_2O_2 in hydrogels containing Mg^{2+}

H_2O_2 stability of hydrogels containing Mg^{2+} was studied to observe the effect of ion charge added into the starting silicate solution. Magnesium was selected to be partially exchanged with the sodium in the silicate at 1:0.2 molar ratio of Mg^{2+}: Na^+ because magnesium ion is divalent and smaller in size than Na^+. Magnesium ion could not be completely replaced with Na^+ because magnesium silicates are not water soluble unlike potassium silicates. The following reaction shows the precipitation reaction of magnesium silicates.

$$2(\equiv SiO^-Na^+) + MgCl_2 \rightarrow 2 \equiv SiO^-Mg^{2+} + 2NaCl \quad (3)$$

The pH of the sol containing Mg^{2+} was decreased using ion exchange resin (Amberlite IR120) which replaces ions with H^+. Figure 4 shows H_2O_2 retention in hydrogels synthesized in the presence (pH=4.6) and absence (pH=8.8) of ion exchange. Previous stability results of hydrogels containing only Na^+ are also included in Figure 4 for comparison. The presence of Mg^{+2} showed improved peroxide stability regardless of the presence of ion exchange. We believe that the reason of this result is more sites available for hydrogen bonding between silanol groups and peroxide molecules due to the smaller size and higher charge of magnesium ions than Na^+.

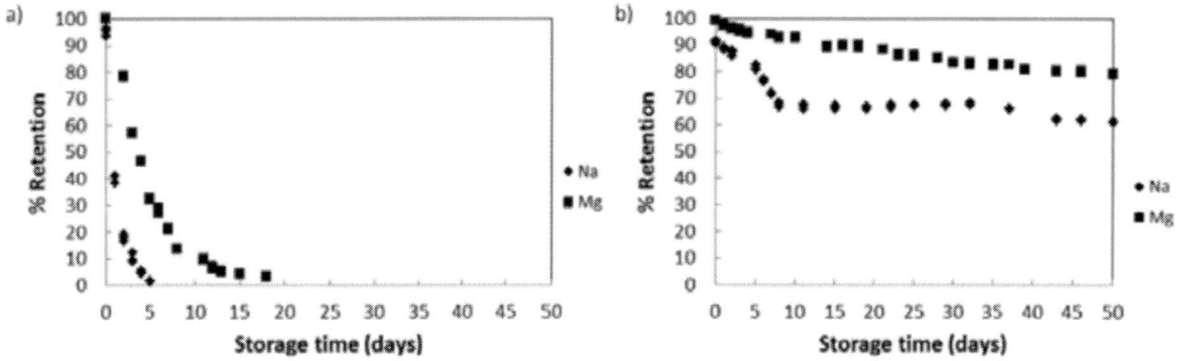

Figure 4. Room temperature stability of H_2O_2 as a function of storage time (a) without ion exchange (pH=8.8) and (b) with ion exchange (pH=4.6)

Morphology and structure of hydrogel

BET and SEM analyses showed that the presence of potassium ion resulted less compact gel structure at constant pH. This is because K^+ has larger atomic size than Na^+ and results larger pores in the hydrogel structure. Figure 5 shows SEM images of samples having different molar ratios of K^+: Na^+ at pH=8.8.

Figure 5. SEM images (at X200,000) of hydrogels for different molar ratios of K^+: Na^+ : a) low, b) equimolar, c) excess (at constant pH=8.8)

Nitrogen sorption studies were carried out to obtain quantitative results of the effect of potassium ion on pore size and surface area of hydrogels. Results are shown in Figure 6. When K^+: Na^+ molar ratio was increased, larger pore sizes were obtained because of the larger atomic size of K^+ (Figure 6a). For each case, small pore sizes were observed as pH decreased because hydrogen ions have capability to bring silica particles together by charge screening effect. In Figure 6b, a decrease in surface area was resulted in the presence of K^+. However, surface area of hydrogels increased at lower pH values as expected due to the decrease in pore size.

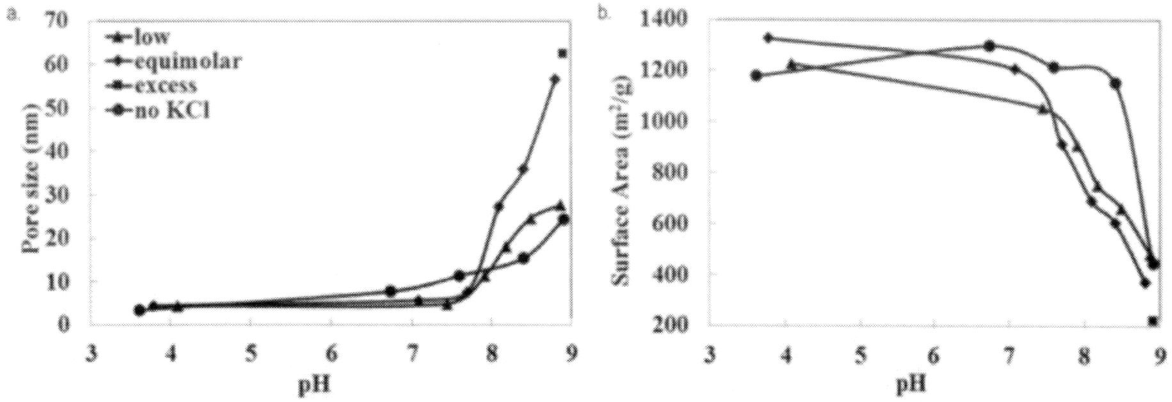

Figure 6. Structural properties of silica hydrogels with changing pH and K^+: Na^+ molar ratio, ▲: low ♦: equimolar ■: excess ●: previous results obtained with unmodified sodium silicate precursor[9]

Conclusions

A significant effect of the size and charge of ions on H_2O_2 stability and the hydrogel structure was observed. Gelation time decreased as K^+: Na^+ ratio were increased in the starting precursor. The stability of encapsulated H_2O_2 was improved with the addition of Mg^{2+} and K^+ ions and decreasing pH of the sol. With the increase in K^+: Na^+ ratio, average pore diameter was found to increase and surface area decreased at constant pH (8.8), therefore the gel became less compact. The higher retention and better stability of H_2O_2 at room temperature was resulted by controlling the ion content (size and charge) and pH. The properties of hydrogels can be tailored to control the gel time and H_2O_2 stability. A detailed study might be required for better understanding of the effect of size and charge of ions on peroxide stability and the hydrogel structure. For example, calcium ions can be included in hydrogel synthesis because the sizes of Ca^{2+} and Na^+ ions are similar whereas it has higher charge like Mg^{2+}.

References

1. Schumb WC, Satterfield CN, Wentworth RL. Hydrogen peroxide. ACS Monograph Series. New York: Reinhold Publishing Corporation, 1955.
2. Sigma Aldrich Product Information Sheet: Hydrogen Peroxide Solution. http://www.sigmaaldrich.com/
3. Colodette JL, Rothenberg S, Dence CW. Factors Effecting Hydrogen Peroxide Stability in the Brightening of Mechanical and Chemimechanical Pulps. J. Pulp. Pap. Sci. 1989; 15: 3–10.
4. Sulieman M, Addy M, MacDonald E, Rees JS. The Effect of Hydrogen Peroxide Concentration on the Outcome of Tooth Whitening: An In Vitro Study. J. Dent. 2004; 32: 295–299.
5. Kim JC, Lee KU, Shin WC, Lee HY, Kim JD, Kim YC, Tae G, Lee KY, Lee SJ, Kim JD. Monoolein Cubic Phases Containing Hydrogen Peroxide. Colloids and Surfaces B: Biointerfaces. 2004; 36: 161–166.

6. Żegliński J, Cabaj A, Strankowski, M, Czerniak J, Haponiuk JT. Silica Xerogel-Hydrogen Peroxide Composites: Their Morphology, Stability, and Antimicrobial Activity. Colloids and Surfaces B: Biointerfaces. 2007; 54: 165−172.
7. Bednarz S, Rys B, Bogdal D. Application of Hydrogen Peroxide Encapsulated in Silica Xerogels to Oxidation Reactions. Molecules. 2012; 17: 8068-8078.
8. Wolanov Y, Shurki A, Prikhodchenko PV, Tripolskaya TA, Novotortsev VM, Pedahzur R, Lev O. Aqueous Stability of Alumina and Silica Perhydrate Hydrogels: Experiments and Computations. Dalton Trans. 2014; 43: 16614.
9. Sudur F, Pleskowicz B, Orbey N. Hydrogen Peroxide Stability in Silica Hydrogels. Ind. Eng. Chem. Res. 2015; 54 (6):1930−1940.
10. Sudur F, Orbey N. Properties of Hydrogen Peroxide Encapsulated in Silica Hydrogels and Xerogels. Ind. Eng. Chem. Res. 2015; 54(44): 11251-11257.
11. Bergna HE, Roberts WO. Colloidal Silica: Fundamentals and Applications. Boca Raton, FL: CRC Press, 2005.
12. Baccile N, Babonneau F, Thomas B, Coradin T. Introducing Ecodesign in Silica Sol-Gel Materials. J. Mater. Chem. 2009; 19: 8537−8559.
13. Brady AP, Brown AG, Huff H. J. The polymerization of aqueous potassium silicate solutions. Colloid Sci.1953; 8(2):252-276.
14. Shroff AV, Shah DL. Grouting Technology in Tunneling and Dam Construction. Rotterdam, Brookfield: Oxford & IBH Publishing Company, 1992.
15. Chun-xi Z. Recent Advances in Water glass Sand Technologies. Shanghai, China: School of Material Science and Engineering, Shanghai Jiaotong University; 2007.
16. 27. Iler RK. The Chemistry of Silica: Solubility, Polymerization, Colloid and Surface Properties and Biochemistry. New York: Wiley, 1979.

472182 UV-Assisted Synthesis of Carbon Nanotube-TiO_2 Nanocomposites for Enhanced Photocatalytic Air Purification

Abstract:

Efficient photocatalysts consisting of carbon nanotubes (CNTs) decorated with TiO2 nanoparticles were synthesized by a simple two-step process that is free of aggressive acid treatments that often introduce defects to CNTs. The novel synthesis process involved UV-assisted oxidation of CNTs using H_2O_2, followed by the deposition of TiO2 via a wet chemical method using TiCl4 as a titanium source. The resulting CNT/TiO2 nanocomposites have been characterized by X-ray diffraction (XRD), transmission electron microscopy (TEM), X-ray photoelectron spectroscopy (XPS), and Raman spectroscopy. XRD results revealed that the crystalline structure of TiO2 is anatase and the TEM results showed the undamaged tubular morphology of CNTs decorated with TiO2 nanoparticles. The XPS data revealed the intimate chemical interactions between the CNTs and TiO2 in the nanocomposites. The expected synergistic effect and efficient separation of photogenerated electron-hole pairs in the synthesized CNT/TiO2 nanocomposites were verified in gas-phase photocatalytic degradation of VOCs.

Copyright © American Institute of Chemical Engineers. All rights reserved.

470129 NOVEL Magnetic Nanocomposite Materials for RAPID Removal of Polychlorinated Biphenyls from Contaminated Water Sources

Abstract:

Magnetic nanocomposite systems able to capture polychlorinated biphenyls (PCBs) in aqueous solution were developed, providing a cost-effective water remediation technique. Two distinct methods were employed to develop these polyphenolic nanocomposite materials, where the polyphenolic moieties were incorporated to create high affinity binding sites for organic pollutants. The first method utilized a surface initiated polymerization of polyphenolic-based crosslinkers and co-monomers on the surface of iron oxide magnetic nanoparticles to create a core-shell nanocomposite. The second method utilized a bulk polymerization method to create macroscale films composed of iron oxide nanoparticles incorporated into a polyphenolic-based polymer matrix, which were then were processed into microparticles. Both methods produce nanocomposite materials that can bind chlorinated organics, can rapidly separate bound organics from contaminated water sources using magnetic decantation, and can use thermal destabilization of the polymer matrix for contaminant release and material regeneration. The polyphenol functionalities used to bind organic pollutants were quercetin multiacrylate (QMA) and curcumin multiacrylate (CMA), which are acrylated forms of the nutrient polyphenols with expected affinity for chlorinated organics. Particles were characterized using transmission electron microscopy (TEM), dynamic light scattering (DLS), Fourier transform infrared spectroscopy (FTIR), thermal gravimetric analysis (TGA) and UV-visible spectroscopy stability analysis. Pollutant binding studies were performed using PCB 126 as a model PCB and chlorinated organic pollutant to determine binding affinity and capacity, and this was quantified using gas chromatography coupled to electron capture detector (GC-ECD). It was demonstrated that the materials effectively bound PCBs in aqueous media.

Copyright © American Institute of Chemical Engineers. All rights reserved.

472163 Influence of Silica-Based NanoparticlesÂ Embedded in Sand Bed Filtration for Cleaning-up Industrial Wastewater

Abstract:

Sand filtration technology is widely used to remove the suspended solids and colloids from the industrial wastewater effluent. However, such standalone technology cannot remove the soluble organic compounds from the industrial wastewater. In the current application, we are targeting to improve not only the removal of suspended solids but also increase the removal efficiency of total organic carbons (TOC) by enhancing the sand surface area via integrating it with functionalized silica-based nanoparticles. These kinds of nanoparticles are environmentally friendly and naturally occurring materials at the macro-level.

In this study, different types of functionalized silica-based nanoparticles were successfully prepared, characterized, and tested for TOC adsorptive removal within batch and continuous adsorption processes. Briefly, the Brunauerâ €"Emmettâ€"Teller (BET) surface area, X-ray diffraction (XRD), Fourier transform infrared (FTIR) spectroscopy and high resolution transmition electron microscopy (HRTEM) were used to confirm the surface area, crystalline domain size, functionality, and structural topology, respectively. The characterization results showed that the prepared nanoparticles are spherical in shape with a crystalline diameter of 10 nm and BET surface area around 180 m^2/g. Adsorption experiments were conducted in a batch mode to investigate the adsorption affinity and capacity of nanoparticles toward TOC removal from a locally supplied industrial wastewater sample. The results indicated that the TOC removal efficiency was improved significantly by 50% using the functionalized sand with nanoparticles at low mass fraction (<10 wt%). Afterwards, the filter bed column, continuous mode, was conducted using the functionalized and the un-functionalized sands. Interestingly enough, the functionalized sand with in-house prepared silica-based nanoparticles showed prolonger breakthrough time (>50 times) in comparison with the un-functionalized sand. These promising results are believed to help the current conventional wastewater treatment methods not only on a lab scale

but also on an industrial level.

Copyright © American Institute of Chemical Engineers. All rights reserved.

464632 Inactivation of *E. coli.* Using a Novel TiO_2 Nanotube Electrode

Abstract:

Electrochemical technology has gained increasing interests for water/wastewater treatment in recent years. In particular, it may find applications in decentralized treatment due to its simplicity, amenability to automation, compact instrumentation, and ease of operation. The performance of electrochemical processes highly depends on the material of electrodes. However, the high cost of electrodes, formation of toxic byproducts, and short electrode lifetime have impeded its wide application. Therefore, development of cost-effective anode materials are needed to support future adoption of electrochemical technologies for environmental applications. In our study, a novel TiO_2 nanotube anode has been fabricated in the lab via simple electrochemical anodization using metallic Ti as the substrate followed by electrochemical hydrogenation/cathodic polarization. The one-dimensional nanostructure is shown in FESEM images. The TiO_2 nanotube anodes with stainless steel as the cathodes were tested for inactivation of a model microbial contaminant *E. Coli.* using a filter-press type cell. Effects of current density applied (0.625 ~ 2.5 mA/cm^2), electrolyte concentration (0.3 mM~3 mM NaCl), and different electrolyte (NaCl, Na_2SO_4, Na_2HPO_4, $NaNO_3$, and $NaHCO_3$) on *E. Coli.* inactivation were investigated. Electrochemically generated oxidizing species were measured as free chlorine. Further, production of reactive oxygen species (•OH, O_2•-, SO_4^{2-}•) were examined using probe compounds. In addition, electro-disinfection experiments were conducted in real water matrices, i.e. samples of surface water, treated surface water before disinfection, and secondary effluent before disinfection collected from local water/ wastewater treatment plants. Formation of inorganic (e.g. ClO_3^-) and organic (THMs) disinfection byproducts were also monitored. Our results have demonstrated that the novel TiO_2 nanotube anode can effectively achieve 5~6 log reduction of E. Coli. with low byproducts formation. The low energy consumption also indicates that such disinfection system may be

operated using solar energy stored in a DC battery for decentralized treatment or as a point-of-use device.

Copyright © American Institute of Chemical Engineers. All rights reserved.

451725 Metal Organic Framework Derived Nanoporous Carbon Â As a Novel Adsorbent for Water Treatment

Abstract:

Metal organic framework derived nanoporous carbon

as a novel adsorbent for water treatment

Zahra Abbasi[a], Ezzatollah Shamsaei[a], Soo Kwan Leong[a], Bradley Ladewig[a,b], Xiwang Zhang[a], Huanting Wang[a*]

[a]Department of Chemical Engineering, Monash University, Clayton, Victoria 3800, Australia

[b]Department of Chemical Engineering, Imperial College London, Exhibition Road, London SW7 2AZ, United Kingdom

Phone: +61 3 9905 3449, Email: Huanting.wang@monash.edu

Nanoporous carbon materials have attracted a great deal of attention because of their potential application in many areas such as adsorption and catalysis. These materials have desirable properties such as high surface area, good chemical and thermal stabilities as well as fast kinetics, which make them suitable for adsorption applications. Porous metal organic frameworks (MOFs) consist of crystalline structures that provide high internal surface areas and tunable porosities which could be considered as starting materials for developing novel carbon based nano composite

adsorbents.

Different types of adsorbents have been studied for the removal of organic pollutants from water and wastewater such as activated carbon, fly ash waste materials and different kind of MOFs. However, their adsorption capacity is low in comparison with nanoporous carbons. Therefore, developing novel carbon based materials with simple synthesis methods for adsorption of pollutants from water is of great importance.

Zeolitic immidazolate frameworks (ZIFs) are a class of MOFs that exhibit remarkable features like high thermal and chemical stability and demonstrate similar pore topologies to zeolites. Among various ZIFs reported, ZIF-8 with the formula of $Zn(2\text{-methylimidazolate})_2$ and with a sodalite-related zeolitic structure has been used for many applications such as adsorption and catalysis. The surface and bulk properties of ZIF-8 can be modified to make it a more competent adsorbent for water treatment purposes. Carbonization of ZIF-8 could be the easiest and most practical modification approach to amend this metal organic frame work's characteristics.

The adsorption capabilities and characteristics of nanoporous carbon particles derived from ZIF-8 were investigated in this study. Direct carbonization of ZIF-8 was used to synthesize carbon nano particles. SEM and TEM images revealed that the ZIF-8 derived carbon carbonized at several temperatures retained the original structure and morphology of ZIF-8. These carbon nano-particles exhibited significantly greater adsorption capacities for the removal of dye from water compared to the ZIF-8. The high adsorption capacity is attributed to the change in surface charge and pore size. The surface functionality and hydrophobicity of the materials were investigated by Raman Spectroscopy, N_2 adsorption-desorption, FTIR, TGA and contact angle measurements. The change in the surface charge of the carbon particles was confirmed by zeta potential of the samples. The dye uptake followed pseudo-second order kinetics and the equilibrium data was best fit to the Langmuir model for the carbon nano-particles. Intra-particle diffusion studies showed that the adsorption process was governed by both the diffusion step and external mass transfer. The ZIF-8 derived nano porous carbons were found to be highly efficient adsorbents for water and waste water treatment.

Copyright © American Institute of Chemical Engineers. All rights reserved.

471195 Study of TL Response of Silver Nanoparticles in Borate Glasses Containing Dy^{3+} and Eu^{3+} Ions for UV and Gamma Dosimetry

Abstract:

Abstract

Dy^{3+} and Eu^{3+} doped the borate glasses samples containing Ag nanoparticles were prepared using a conventional melt-quenching method. The synthesized materials were characterized by scanning electron microscopy (SEM) and their morphology and sized distributions were determined. The photoluminescence (PL) spectra as well as UV-Vis absorbance were also analyzed and show the presence of the rare-earth in the matrix. The thermoluminescence (TL) response of Dy^{3+} doped lithium borate glasses irradiated with ultraviolet (UV) radiation was studied in detail before and after doping with silver.

Copyright © American Institute of Chemical Engineers. All rights reserved.

471313 Effects in Morphology and Thermoluminescent Characteristics of Lif Crystals Synthesized By Using Nonionic and Cationic Surfactants

Abstract:

LiF crystals doped with Ce^{3+} and Mg were prepared using co-precipitation, using as solvent a mixture of ethanol and water in presence of Nonionic (PVP and Pluronic f-127) and cationic (CTAB) surfactants. The LiF crystals were characterized by scanning electron microscopy (SEM) obtaining average sizes of 100 nm, a single cubic crystalline phase was determined by XRD, the PL emission spectra shows two bands characteristics of Ce^{3+} and UV-Vis absorbance shows the presence of the rare-earth in the LiF crystals. The TL response to both X-ray and gamma from a Cs-137 source irradiation was studied.

Copyright © American Institute of Chemical Engineers. All rights reserved.

470030 Thermoluminescent Dosimetric Analysis of Ag and Cu Doped LiB3O5

Abstract:

Methods

Production

Synthesis of Ag and Cu doped LiB_3O_5 crystals was carried out by precipitation, using water as a solvent. Impure crystals were produced at a synthesis temperature of 70 °C. A 5g of Ag and Cu doped LiB_3O_5 precipitate was obtained upon evaporation of water in the solution. The precipitates were calcined at a temperature of 750 °C for one hour. The calcined materials were grounded finely and subjected to severalcharacterizations.

Characterization

SEM analyses were performed for the doped and undoped samples of LiB_3O_5. The XRD properties of the samples were obtained. The UV-Vis absorbance spectra of the samples were obtained at room temperature. Spectra were recorded between 200 and 1200 nm. The photoluminescence (PL) measurements were carried out. The glow curves of the samples has been obtained using TLD reader.

Results

The XRD patterns obtained for the samples showedthat production of LiB_3O_5 with orthorhombic structure was successful since allof the main peaks of LiB_3O_5 coincide with the JCPDS data. It was seen that the dopant added did

not interfere withthe crystal structure. As it can easily be detected addition of the dopants didnot have any significant effect on the crystal structure of LiB_3O_5 produced.

SEM was carried out to check the morphologies of the samples.According to the SEM images, the samples produced by the co-precipitation method shows the agglomerated surface.

The UV-Vis absorption spectra for the samples showed that the maximum absorbance was observed at 250nm. The result showed that a clear monotonic and increasing dependence of the intensity of the absorbance as a function of dopants.

The dosimetric properties of LiB_3O_5 doped with Ag and Cu were determined by using HarshawTLD Reader Model 3500. The heating rate was chosen as 10°C/s was usedfrom 50 to 350°C. The examined dosimeters were exposed to X-rays at room temperature. The sample amount was 0.5g. The samples have shown the TL response. The pattern of the glow curves changed with the change of dopants.

Conclusions

The TL response of LiB_3O_5 has been compared with the commercial TLD100. The observation concludes that the doped LiB_3O_5 has the TL behavior. The glow curves of the LiB_3O_5 shows different shape with the change of dopants.

Copyright © American Institute of Chemical Engineers. All rights reserved.

469662 Effect of Thermal Treatment on the Characteristics of PES/PVA Nanocomposite Membranes Modified with TiO2 Nanoparticles :a Comparitive Study Between 1-Step and 2-Step Thermal Treatment

Abstract:

PES/PVA thin film composites were modified with TiO2 nanoparticles via dip coating method. Thermal treatment of total modified membranes plays a semi-"catalytic role" in coating process. Hence, 4-temprature levels were selected and investigated, noticing that there is an important restriction due to the stability limitation of polymers under thermal treatment process. This phenomenon also leads to the concept of 2-step treatment. 2-step thermal treatment is a more moderate process which leads to a more stable structure.

Polyethersulfone (PES) ultrafiltration membrane was prepared via phase inversion induced by immersion precipitation. Poly(vinyl-alcohol) (PVA) (1.5 wt%) dissolved in distillated water was coated on the surface of PES barrier layer with dip coating technique. The PVA-coated membranes were then immersed in the cross-linking solution containing 5 wt.% of GA and 0.5 wt.% of H2SO4 for 2min at room temperature in order to reduce the membrane swelling by chemical cross-linking. H2SO4 was used as catalyst of reaction. TiO2 nanoparticles were dispersed in distillated water by sonication for 1hr and stirred vigorously for 30 min. the PVA coated membranes were immersed in 0.2 wt% TiO2 solution for 15 min. the last step is thermal treatment which in this study, four level of temperature (85,95.105.115) were investigated. All modified membranes were thermal treated for 3 min.

The performance of modified membranes was tested with filtration of yeast industrial wastewater(COD=11500 mg/L). Permeate flux and COD of four modified membranes have tested. According to the results, the PES/PVA/TiO2 membrane that was thermal treated in 105 had the best output.

Copyright © American Institute of Chemical Engineers. All rights reserved.

469788 Synthesis, Structural and Morphological Characterization of Cu and Ag-Doped $Li_2B_4O_7$

Abstract:

Method:

In this paper the lithium tetra borate (LTB) was produced by water/solution assisted synthesis method. After the synthesis, metals (doping with Copper (Cu) and Silver (Ag)) were doped into lithium tetra borate ($Li_2B_4O_7$) to enhance its thermo luminescent properties. The heating temperature was 750°C for 2 hr as well as 150°C for 1hr the sample produced by water solution method with 0.1% Cu and 0.1% Ag. The pellets were irradiating by the X-ray.

Result:

The identification and characteristics of undoped and doped $Li_2B_4O_7$ were determined by X-ray diffraction (XRD), Scanning electron microscope (SEM), Photoluminescence (PL), Ultra violet visible spectroscopy (UV Vis). The chemical compositition and its morphologies of the obtained LTB and LTB: Cu, Ag was confirmed by XRD and SEM results. The most intense peak of the XRD pattern of the lithium tetraborate sample is verified by the reference data of the lithium tetra borate with tetragonal structure. The glow curves were obtained by thermo luminescence (TL) reader. The glow curves of the pellets exposed to different doses exhibited a clear response to X-ray irradiation, which is a significant evidence that this material is a candidate for radiation dosimetry. Finally this energy response was also compared with TLD100. The comparison of the results showed that the obtained lithium tetra borate have good thermo luminescence (TL) dose response.

Copyright © American Institute of Chemical Engineers. All rights reserved.

Enhancement of Heat Transfer Coefficient by Using Fe_2O_3 –Water Nanofluids

N.I. Zouli[1], Muthanna Al-Dahhan[1]
[1]Missouri University of Science and Technology, Rolla, MO, 65409
Email: niz8vb@mst.edu

Introduction

Research efforts have been made to increase the heat transfer rate, energy, and efficiency. Some of these efforts include the application of fins, compact heat exchangers, microchannel, and many others. Recently, nanofluids have been introduced to enhance the heat transfer rate of the heat transfer fluids by adding a certain concentration of nanoparticles into the heat transfer media. Most of the work that has been published in the literature focuses on measuring the thermal conductivity of the nanofluid or measuring the overall heat transfer coefficient. Among the efforts for enhancement of heat transfer the application of additives to liquids is noticeable (Bergles, 1973; Ahuja, 1975). Choi (1995) was the first who studied the enhancement of the nanoparticles on the heat transfer coefficient and showed that the nanoparticles could considerably affect the thermal conductivity of the nanofluid to increase. Since Choi is work (Choi,1995), various studies have been conducted on nanofluids and these studies shown that nanofluids increase heat transfer properties and stability while reducing costs, clogging, and energy required to pump the fluid. Heris et al. (2006) found that the heat transfer coefficient increased as the volume concentration of the nanofluids increased when the fluid was composed of water and metal oxides.

In the present study, effects of Fe_2O_3/water on the nanofluid's heat transfer and thermal conductivity will be discussed. The experimental investigations are done to study heat transfer performance of the nanofluid at the entrance region under laminar flow condition varying the particle size and volume concentration of nanoparticles.

Experimental setup and methodology

Schematic of the experimental setup used in the current study is shown in Figure1. It consists of a flow loop made of copper material, a heating unit, a measuring and control unit. A straight copper tube with 190.5 mm length, 25.4 ± 0.02 mm inner diameter, and 31 ± 0.05 mm outer diameter was used as the test section. The heat transfer probe comprises of a heat flux sensor (0.15 in x0.08 in) with a fast response time of about 0.02 sec and a thermal resistance of 0.003 °F/BTU/ft2hr. The sensor was placed in a slit mounted inside the test section wall flashed to the wall without disturbing the flow. The probe measures the heat flux and the surface temperature. In addition, two thermocouples (type T) were mounted inside the test section to measure the bulk temperature. The sensor was placed on the middle distance between the two thermocouples.

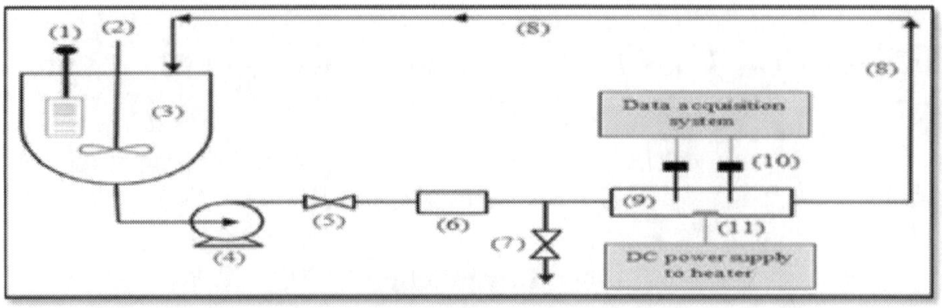

Figure 1. Schematic of experimental setup components: (1) Heater, (2) Agitator, (3) Reservoir tank, (4) Pump, (5) Needle valve, (6) Flow meter, (7) Drainage valve, (8) Copper tube, (9) Test section of copper tube, (10) Thermocouples, (11) Heat flux foil sensor.

Heat flux, temperature of the sensor surface, and the bulk media temperature are directly measured and recorded using Data acquisition system. Further they are used to estimate local instantaneous and averaged heat transfer coefficient using eqn (1) and (2):

$$h_i = \frac{q_i}{T_{si} - T_{bi}} \; ; \; \Delta T_i = T_{si} - T_{bi} \qquad (1)$$

$$h = \frac{1}{n} \sum_{i=1}^{n} \frac{q_i}{T_{si} - T_{bi}} = \frac{1}{n} \sum_{i=1}^{n} h_i \qquad (2)$$

where h_i: local instantaneous heat transfer coefficient (kW/m^2.K)
q_i: instantaneous heat flux measured by the sensor (kW/m^2)
T_{si}: instantaneous surface temperature of the probe (K)
T_{bi}: instantaneous bulk temperature of the media (K)

Results and discussion

The thermal conductivity of Fe_2O_3 water based nanofluids was measured at different concentrations. From Figure 2 an enhancement of 22.35% in the thermal conductivity value has been achieved with 0.09% volume concentration of Fe_2O_3 nanoparticles in water with an average diameter of 20 nm. As can be seen from Figure 2, the thermal conductivity enhancement increases linearly with increased particle volume concentration.

The concentrations for each test run were increased from 0.02-0.09% by volume and run at a constant flow rate of 6 L/min. The experimental results were obtained with 20 nm particles. And it was found that the average increase in the heat transfer coefficient was between 5.95-18.34% with the highest increase being 73.7% at 0.09% volume concentration at 5 mins into the process. It was observed that as the concentration of the nanofluid was increased so did the heat transfer coefficient. The enhancement in heat transfer coefficient was calculated by following;

$$[(h_{nf} - h_{bf}) / h_{bf}] * 100\% \qquad (3)$$

h_{nf} = heat transfer rate of nanofluid
h_{bf} = heat transfer rate of base fluid

A typical result for 20 nm particles can be seen in Figure 3. Overall the results show that the local heat transfer coefficient increases with increasing the size and the volume fractions of the Fe_2O_3 nanoparticles compared to water. The effect of Reynolds number on Nusselt number is shown in Figure 4. It is seen that the measured Nusselt numbers for nanofluids are higher than those of base fluid, and they increase remarkably with Reynolds number. The observed enhancement of the Nusselt number could be due to the suppression of the boundary layer, the viscosity of nanofluids as well as the dispersion of the nanoparticles.

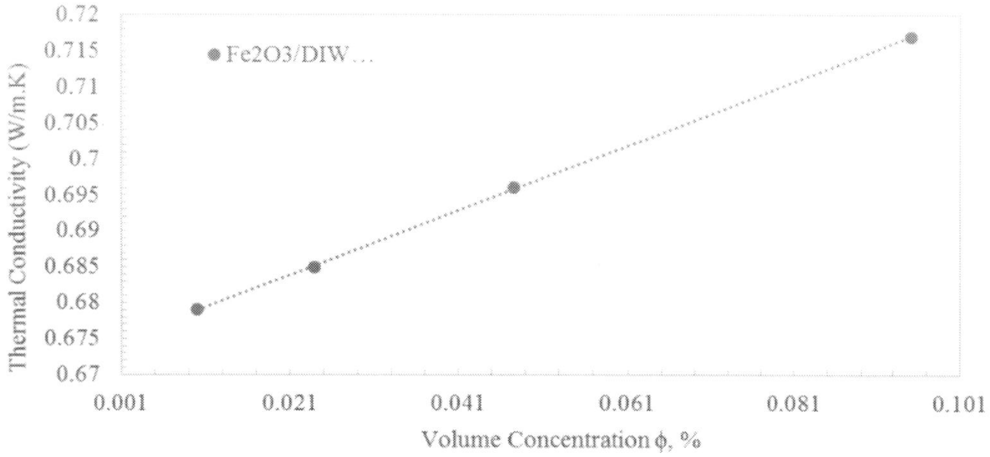

Figure 2. Thermal conductivity of Fe_2O_3 water based nanofluids at different volume fractions (0 and 0.09 %).

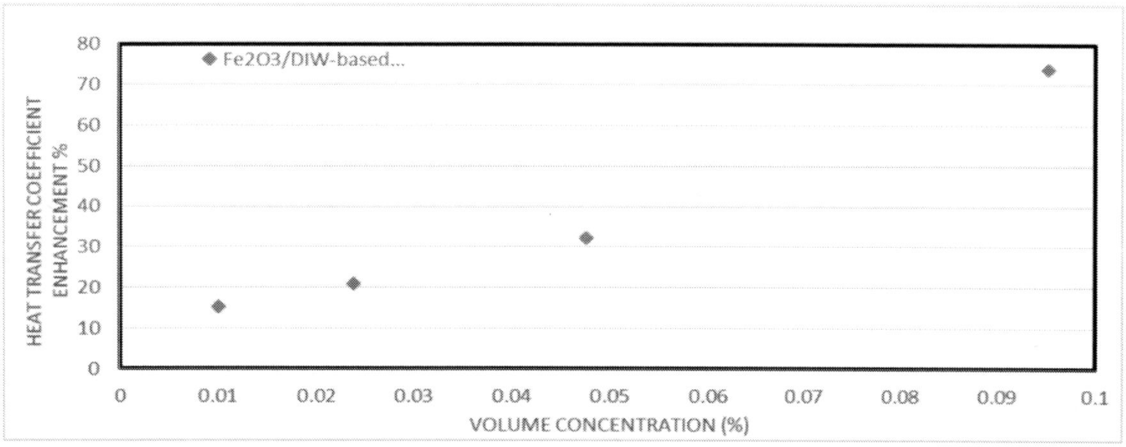

Figure 3. Variation of heat transfer coefficient with different volume concentration of Fe_2O_3 water based nanofluids.

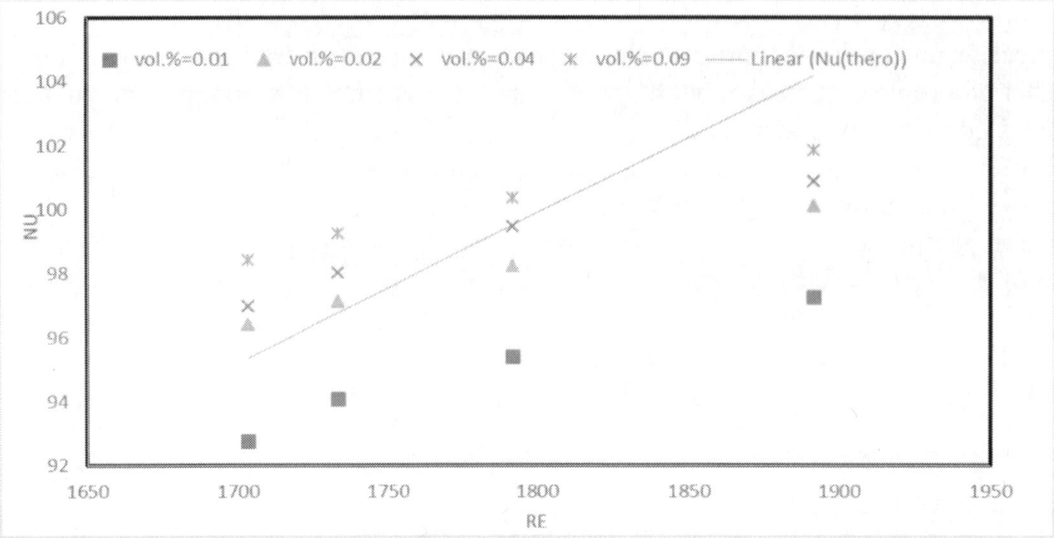

Figure 4. Effect of Reynolds number and Fe_2O_3 nanoparticle concentration on Nusselt number

Conclusions

The paper is mainly concerned with the heat transfer of nanofluids made of water and Fe_2O_3 nanoparticles in the laminar flow regime. It was found that use of aqueous Fe_2O_3 nanoparticles can significantly enhance heat transfer coefficient in laminar flow regime, and the enhancement increases with Reynolds number, as well as particle concentration under the conditions of this work. The enhancement of the heat transfer coefficient could not be only credited to the enhancement of the effective thermal conductivity. A reason for the enhancement, Particle passage which lead to reduce the thermal boundary layer thickness and a non-uniform distribution of thermal conductivity.

Acknowledgments

I would like to acknowledge the financial support provided by the government of Saudi Arabia. I would also like to thank Prof. Al-Dahhan for his supervision and support which made this work possible.

References

1. Brooke, A., Kendrick, D., Meeraus, A. A. (1992). GAMS- A User's Guide (Release 2.25). *The Scientific Press*. San Francisco, CA.

2. Gooding, W. B., Pekny, J. F., McCroskey, P. S. (1994). Enumerative Approaches to Parallel Flowshop Scheduling via Problem Transformation. *Comput. Chem. Eng.*, *18*, 909.

3. OSL. (1991). *OSL, Guide and Reference (release 2)*, IBM: Kingston, NY.

4. Zhang, X., Sargent, R. W. H. (1994). The Optimal Operation of Mixed Production Facilities - A General Formulation and some Approaches for the Solution. *In Proceedings of the 5th International Symposium on Process Systems Engineering*. Kyongju, Korea, 171.

5. A.E. Bergles, Recent development in convective heat transfer augmentation, Appl. Mech. Rev. 26 (1973) 675–682.

6. A.S. Ahuja, Augmentation of heat transport in laminar flow of polystyrene suspension: I. Experimental and results, J. Appl. Phys. 46 (8) (1975) 3408–3416.

7. Choi, S.U.S. Enhancing thermal conductivity of fluids with nanoparticles. in American Society of Mechanical Engineers, Fluids Engineering Division (Publication) FED. 1995.

8. Heris, S.Z., S.G. Etemad, and M.N. Esfahany, *Experimental investigation of oxide nanofluids laminar flow convective heat transfer.* International Communications in Heat and Mass Transfer, 2006. **33**(4): p. 529-535.

447375 Eco-Friendly Dyeing of Electrospun Cellulose Nanofibers with Reactive Dye Using Ultrasonic Energy

Abstract:

Nanofibers have created enormous opportunities and challenges for the textile and apparel industry, moreover the innovative methods for dyeing of nanofibers have recently gained more attention. Electrospun cellulose nanofibers are increasing significant interest in textile industry in view of its incomparable properties. However, the dyeing of nanofibers faces problem of lower color yield due to the higher surface area of nanofibers. In this study electrospun cellulose nanofiber dyed with using cold pad batch dyeing, a well-known and ecofriendly method of dyeing for cotton base fibers. The dyeing of cellulose nanofibers was carried out with CI Reactive Violet 5 by conventional and ultrasonic (US) method. Scanning electron microscopy and Fourier transform infrared spectroscopy were used to investigate morphology and chemical structure of electrospun cellulose nanofibers before and after dyeing and the color strength (K/S value) and color fastness of dyed nanofibers also assessed.

The results indicates that the use of ultrasonic energy not only shortens the batching time but the color strength (K/S),color fixation and color fastness also considerably increased ,the K/S value of dyed samples enhanced around 5-8% compared with conventional method. Although, washing fastness of sample increased up to 25%, but, color fastness to light was the same as conventional method.

The results proved that, the morphology of dyed cellulose nanofibers with both dyeing method were smooth with structural uniformity and remained unaffected after being dyed.

Copyright © American Institute of Chemical Engineers. All rights reserved.

452910 Combined Quartz Crystal Microbalance with Dissipation and Generalized Ellipsometry to Characterize the Deposition of Titanium Dioxide Nanoparticles on Model Rough Surfaces

Abstract:

Titanium dioxide nanoparticles (TiO_2NPs) are the most extensively manufactured engineered nanomaterials, and understanding of their interactions with soil and sandy aquifer materials is necessary for effective natural resource management [1]. Quartz crystal microbalance with dissipation monitoring (QCM-D) has been widely used to study TiO_2NPs at water-metal oxide interfaces as model systems for soils and aquifers. However, natural sands and sediments exhibit various degrees of surface roughness, and the effects of surface roughness are not accounted for in current QCM-D studies of nanoparticle adsorption, which use the Voigt-Kelvin model [2].

A new mechanical model that considers both viscoelastic and surface roughness effects [3] is described to analyze QCM-D measurements. Obtained parameters include adsorbate areal mass density, adsorbate viscoelastic properties, the vertical roughness length, and the permeability length. Silicon slanted columnar thin films (SCTFs) were deposited onto QCM-D sensors by electron-beam glancing angle deposition and conformably coated with an alumina film via atomic layer deposition to act as model rough surfaces. The adsorption of polyacrylate-stabilized TiO_2NPs onto flat surfaces and the SCTFs was measured by QCM-D and generalized ellipsometry (GE), an optical reference technique.

We report that the surface-roughness-modified mechanical model excludes water that is retained by the SCTF surface roughness and instead considers the surface mass density of TiO_2NPs and water coupled to the TiO_2NPs [4]. GE is also sensitive to the attachment of the TiO_2NPs. When the results of both techniques are combined, new information, the hydration of the polyacrylate-stabilized TiO_2NPs, can be evaluated.

[1] B. Kim et al. J. Environ. Monit., **14**, 1128 (2012).

[2] M.V. Voinova et al. Phys. Scr., **59**, 391 (1999).

[3] L. Daikhin et al. Anal. Chem., **74**, 554 (2002).

[4] N. Kananizadeh et al. J. Hazard. Mater., In press. dx.doi.org/10.1016/j.jhazmat.2016.03.048

Copyright © American Institute of Chemical Engineers. All rights reserved.

AUTHOR INDEX